AN INTRODUCTI

GRAND CANYON | FOSSILS

by Dave Thayer

GRAND CANYON
ASSOCIATION

Grand Canyon Association

P.O. Box 399

Grand Canyon, AZ 86023-0399

(800) 858-2808

www.grandcanyon.org

Edited by Todd Berger, James Buckley Jr., and Beth Adelman

Printed in China

First Edition

13   12   11   10   09   1   2   3   4   5

Cover photograph: Michael Quinn/NPS

Title page photograph: Michael Quinn/NPS

Pages 2-3 photograph: iStock

Library of Congress Cataloging-in-Publication Data

Thayer, Dave.
  An introduction to Grand Canyon fossils / by Dave Thayer. -- 1st ed.
      p. cm.
  Includes bibliographical references and index.
  ISBN 978-0-938216-95-7 (alk. paper)
  1. Fossils--Arizona--Grand Canyon National Park. 2. Paleontology--Arizona--Grand Canyon National Park. 3. Geology--Arizona--Grand Canyon National Park. 4. Geology, Stratigraphic. I. Title.
QE747.A7.T53 2009
560.9791'32--dc22

                                          2008037041

Acknowledgments

This book is dedicated to my dear wife, Dora Thayer. —D. T.

CONTENTS

Prologue: A DISCOVERY

November 2002, and a chill wind whipped the junipers. My fingers scraped across icy rocks as I groped from fossil to fossil on a series of bedrock ledges in the upper Kaibab Formation along the Hermit Trail at Grand Canyon National Park. Dozens of marine fossils and fragments protruded from the rock face. I recognized crinoid columnals, types of fossils that look like little beads pierced by tiny holes. The wall also harbored brachiopod shells, mostly fragments, but I saw a few whole ones: *Derbyia, Peniculauris,* and *Meekella.* The sponge *Actinocoelia,* checkered with tan and white, decorated the centers of round, four-inch chert nodules.

And suddenly, there it was—what I had sought in vain so many times in this uppermost Grand Canyon layer: a trilobite fossil. Not just any trilobite, but one from the Permian Period—among the last of its kind to live on Earth. Putting my nose to the rock with

An assemblage of crinoid fossil columnals in Grand Canyon's Kaibab Formation

my little magnifier half an inch from the dime-size fossil, I saw that it was a full tail section of *Ditomopyge,* an extremely rare find.

It is illegal to collect any items, including fossils and rocks, in a national park. So, instead of taking the trilobite, I described it. Placing a centimeter scale beside the precious fossil, I took several photos. I drew a pencil sketch, making sure to include the tiny, smooth bumps that rimmed the tail. I recorded the exact location in my notebook and on my GPS receiver. Only after all that was done, I looked up and down the trail and, seeing no one, let out a triumphant shout and listened for the echo.

Although 270 million years separated me from this tiny creature, I could still see and touch its remains. Generations of geologists have collected data here, and, using their insights, I could imagine a day in the trilobite's life.

Today, the Kaibab Plateau rises more than 7,000 feet above sea level, but 270 million years ago, the rock layers that make up the plateau were beneath and part of the last Permian sea. As it receded, the sea left behind its sands and carbonate muds. Living in this remaining muck, the one-inch-long trilobite looked rather like a horseshoe crab. Rhythmic movements propelled it in search of food. As the trilobite touched bottom, its tiny legs made two rows of dimpled tracks. Then, the creature disappeared into the mud, becoming buried and lost in time—until today.

My mind came back to the present. How might I connect to this trilobite's impossibly distant world? With only a 10-power magnifier, I am peering deeply into time, long before mammals or dinosaurs. This is the key to my fascination with Grand Canyon geology: Like so many others who peer into rocks or through telescopes, I am a time junkie. Nothing quite satisfies like a journey to long ago.

I learned to pick through shreds of time, assembling bits and pieces of ancient life and landscapes. It was not only animals and plants that evolved over millions of years, but the habitats they occupied. Imperceptibly, oceans became shallows, then beaches, then coastal deserts. Animals and plants would follow these shifting niches or die. Then, in cycles slow beyond human patience, the sea crept back. The layers of the Grand Canyon formed.

From the South Rim's Lipan Point, one can see evidence that, more than a dozen times, seas visited this area and then retreated, leaving their sediments. Some seas left several sediment layers, others only one, and others none at all. Some seas were like fugitives walking backward and dragging branches to conceal their footprints, the oceans' retreating wave fronts erasing any deposits and organisms they may have harbored. Other seas left plenty of evidence: thick layers of fossil-filled limestones and shales. When the seas retreated, land deposits formed—swamps, deltas, and deserts.

Where can you look in the Grand Canyon for fossils? A ranger walk can help. From the South Rim, you're guided to limestone beds in the Kaibab Formation. You'll find fossils along the trails, especially in the canyon. It helps to know in advance what you are looking for by studying fossil books. The best place to look for fossils is in shale or limestone beds. Take along a camera and a notepad, and keep your eyes peeled. Snap photos and take notes. Locating a fossil can take you into a seemingly lost world. Resting where it has for 270 million years, my trilobite is still there; I visit it from time to time.

INTRODUCTION TO FOSSILS AND PALEONTOLOGY

Before we climb into the Grand Canyon, let's take a broader look at how fossils are formed and found. Also discussed are some terms and timelines that are important to an understanding of fossils in general and ancient life in the Grand Canyon in particular.

A Succession of Ecologies

Most of the animals and plants of ancient times are extinct today. Wouldn't it be fascinating to visit many different past environments? Had we lived 542 million years ago, at the time of the earliest trilobites, we would have experienced a planet quite different from today's. We cannot travel there, so we must search for clues in the rocks. Among these clues are fossils, which, along with their enclosing rocks, give us portraits of bygone oceans, deserts, forests, mountains, swamps, reefs, and shorelines.

Meekella, a brachiopod fossil found in the Kaibab Formation rock layer at Grand Canyon

The word "ecology" comes from the Greek word *oikos*, which means "house." The suffix "paleo" is derived from the Greek word *palaio*, which means "ancient." Therefore, paleoecology is the study of the ancient life and environments of our home, planet Earth. Paleoecology teaches about past global warmings and ice ages so that we can better monitor our present climate and understand what is "normal." Many paleoecologists will admit that they took up the subject because it fascinates them. Hunting for fossils and trying to understand what they were like when they were alive is like solving a puzzle, and paleoecologists spend many a late evening groping over stones until it is too dark to see. As they rest from their day on the rocks, they imagine the sun's warmth on a Paleozoic amphibian's cheek, or wind blowing across the rolling surface of a sea that no human ever saw.

Paleoecology also helps in the search for fossils. Look out at the sedimentary layers or strata of the Grand Canyon, so different in color and texture. All were formed on flat ground near sea level, before the Grand Canyon's plateaus uplifted. This is why the layers look horizontal. Almost every layer contains fossils, each representing an ancient ecosystem—organisms interacting with one another and with their environments.

Each sedimentary stratum represents the surface of the Earth at some past interval. Even a single layer may represent many millions of years. A drive to the canyon's South Rim illustrates this: forest to savannah to grassland and back again to forest in a matter of 50 miles. Through time and drought, the grasslands might become desert and the forests grasslands. All these changes in time and space are recorded in the rocks. Grand Canyon is one of the most complete such records in the world.

Coconino Sandstone sits atop the darker red rock of the Hermit Formation. These sedimentary strata contain clues to their differing environments, including fossils of the animals and plants that once lived there.

Fossils and Paleontology

Fossils are the remains or traces of ancient life. Even ancient tracks, trails, and burrows are fossils. The scientists who dedicate themselves to studying fossils are called paleontologists. As a rule, a paleontologist is concerned with fossils older than human civilization, leaving cultural sites to archaeologists.

Paleontology has many rewarding disciplines; only a few paleontologists dig up dinosaurs. For example, there's the functional morphologist, a paleontologist who discovers a fossil organism's physical capabilities by analyzing its skeletal structure. It was a functional morphologist who hypothesized that *Tyrannosaurus rex* was probably a scavenger.

The biostratigrapher studies rock layers and correlates them with other rock layers worldwide, comparing fossils that occurred during a certain interval of geologic time in several places. Ichnologists study ancient tracks and trails. Other researchers study evolution, mass extinction, past climates, paleobotany, or fossil data regarding plate tectonics. And these are only a few of the specialized areas of study available to paleontologists.

Grand Canyon National Park provides innumerable sites for professional paleontologists with permits from the National Park Service to use their skills and tools to delve into the past, searching the fossil record. Amateur fossil hunters can use their eyes and cameras to find and record fossils in the park.

It takes specific environmental conditions for an organism to become a fossil. The most important are absence of oxygen (which would cause decomposition instead of fossilization), lack of scavengers, a low-impact environment where the organism won't be ground to bits, and a scarcity of soil-churners and decomposers such as worms, insects, and bacteria. Rapid burial can provide most of these conditions. Organisms submerged in anaerobic water, which is stagnant and oxygen free, are conducive to fossilization because scavengers and decomposers cannot live in it. After burial, groundwater can soak into organic remains, introducing mineral matter and causing petrification.

When an organism becomes fossilized, the original hard material—such as woody material, a skeleton, or a shell—may be preserved. More often, the original material is petrified by soaking in mineral-laden groundwater. The organic remains are mineralized. Mineralization can preserve even the microscopic structure of a plant or animal, but usually the soft tissue is lost.

Fortunately, many sea creatures have hard shells that fossilize readily. Such fossils are often encased in solid stone. When a fossil is preserved by durable rock materials such as chert, the surrounding rock can wear away, leaving a fossil in bas-relief. Many of the fossils of the Kaibab Formation, the Grand Canyon's top layer, exhibit differential erosion of contrastingly hard chert and soft limestone.

Transgression and Regression

When a sea covers part or all of a continent, it's called a transgression. When the sea retreats, it's a regression. The causes of transgressions and regressions may be either changes in sea level or a raising or lowering of the land. For example, when glaciers form and lock up water in ice, sea level goes down. Alternatively, a continental shoreline might subside due to the weight of sediment, and this subsidence causes the sea to encroach.

Geologic Time

Geologic time is like the time on a clock—just lots more of it! More than 1,700 million (or 1.7 billion) years ago, the moon shone on mountains in what is now the Grand Canyon region. No organisms saw

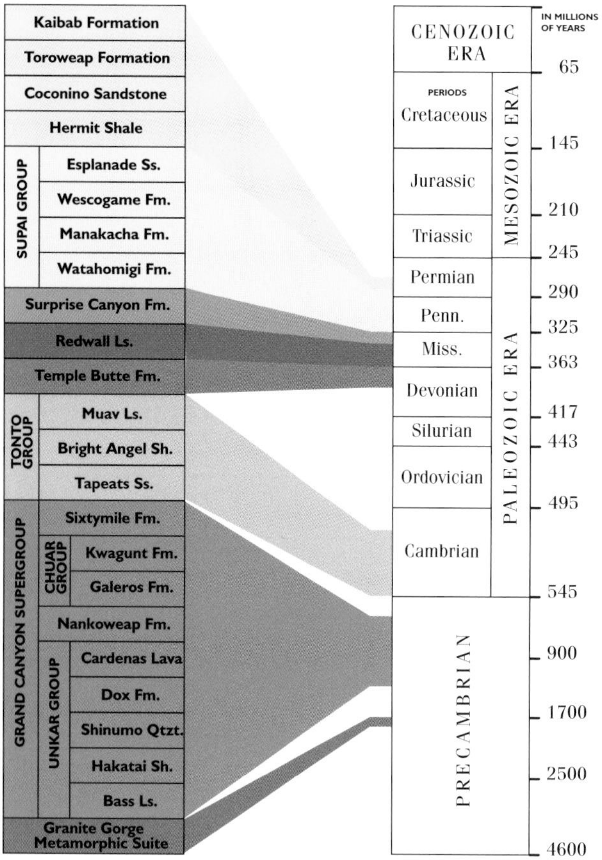
Kaibab Formation
Toroweap Formation
Coconino Sandstone
Hermit Shale
SUPAI GROUP
Esplanade Ss.
Wescogame Fm.
Manakacha Fm.
Watahomigi Fm.
Surprise Canyon Fm.
Redwall Ls.
Temple Butte Fm.
TONTO GROUP
Muav Ls.
Bright Angel Sh.
Tapeats Ss.
GRAND CANYON SUPERGROUP
Sixtymile Fm.
CHUAR GROUP
Kwagunt Fm.
Galeros Fm.
Nankoweap Fm.
UNKAR GROUP
Cardenas Lava
Dox Fm.
Shinumo Qtzt.
Hakatai Sh.
Bass Ls.
Granite Gorge Metamorphic Suite
CENOZOIC ERA
PERIODS
Cretaceous
Jurassic
Triassic
MESOZOIC ERA
Permian
Penn.
Miss.
Devonian
Silurian
Ordovician
Cambrian
PALEOZOIC ERA
PRECAMBRIAN
IN MILLIONS OF YEARS
65
145
210
245
290
325
363
417
443
495
545
900
1700
2500
4600

This chart shows the different time periods into which geologists and paleontologists divide time, along with the major rock layers in the Grand Canyon. Many rock layers take their names from local landmarks.

the mountains—there were no eyes yet. No one can imagine the number of slowly expanding dawns or painted sunsets. Geologic time charts give us the illusion of understanding, but it is only an illusion. We cannot imagine our human ancestors' lives during each of the 50,000 generations that cover only 1 million years.

The geologic time chart, by giving names for geological periods, can be useful in communicating the course of the fossil record. By referring to a time period as "Cambrian," for example, all those who work with geologic time will know what is meant—the time of the trilobite and the first large blossoming of life. If, instead, you refer to only the number, in millions of years ago, you might soon find yourself confused, because this figure will change by a small amount as geologists gather more data to pin down the beginning and end of any geologic period. A memorized date will be out of date in no time at all!

To call a geologic period by its name is to use relative dating. Even though the numbers change, the order in which events occurred does not. Saying an event occurred in the Pennsylvanian Period makes it clear that it came after the Mississippian Period and before the Permian. Relative dating is independent of absolute age, and requires only that we determine the chronological order of events. Conversely, absolute dating means applying a number, an exact age or age range, to an event—for example, 100 million years ago, plus or minus 2 million years. To determine the absolute age, we use radiometric dating. This method measures the decay of radioactive uranium, potassium, or other elements, giving us a geologic elapsed-time stopwatch.

There are three types of rocks. An igneous rock is any rock that melted and hardened again, such as lava. A metamorphic rock is any rock that has changed in texture or mineralogy as a result of heat and pressure (but not enough heat to melt it, or it would be igneous). A sedimentary rock is formed by erosion of dissolved or granular sediment. Absolute dating is usually possible only on igneous and metamorphic rocks, so scientists rely mostly on relative dating, using fossils, for sedimentary rocks.

Using terms from the geologic time chart, the oldest rocks in

Paleontologists can travel backward in time by looking at the layers of rock revealed as the Colorado River slowly carved the Grand Canyon.

The Grand Canyon's Hermit Formation is shale and mudstone, rocks well-suited to preserving this dragonfly wing fossil.

Grand Canyon are early Proterozoic. These are found in the Inner Gorge. Most of the rest of Grand Canyon rocks are Paleozoic in age, coming before the Mesozoic Era. The Colorado River cut the Grand Canyon much later, during the Cenozoic Era, the Age of Mammals.

An important concept in paleontology is the Law of Faunal Succession. It states that a species never appears at two different times, discontinuously, in the fossil record. In other words, extinction is permanent. Using that idea, if we find a certain assemblage of fossils in one area and the same fossils in a far-removed area, we can be sure both assemblages were formed during the same interval of time. Their enclosing rocks must also be of the same age, meaning they are correlated—for example, they may all belong to the Devonian Period. Correlation of sedimentary rocks from place to place is one of the puzzles of geology that can be solved by analysis of fossils. Since radiometric dates cannot be determined for most sedimentary rocks, fossil correlation is invaluable.

Unconformities: Gaps in the Rock Record

During the Grand Canyon region's history, long periods of time went by during which sediment layers did not form, or they did form but were eroded away. This resulted in gaps in the rock and fossil record, known as unconformities. For example, the Tonto Sea encroached over what is now Arizona about 525 million years ago

Geologic time can be rather daunting to consider. As one way of illustrating its sweep, the key geologic events of the Grand Canyon can be figuratively compressed. In fact, imagine all of Earth's 4.54 billion years compressed into one year. On this scale, Earth's earliest known rocks formed in late February. The earliest evidence of microscopic life on Earth formed in March. Mighty mountains rose at the future site of the Grand Canyon in mid-July. The roots of these mountains, the Vishnu basement rocks, are exposed in the Inner Gorge layers. The mountains stretched from Wyoming to Mexico, using modern geographical terms. Rain and other erosive processes washed the mountains down to sea level by late September, about the time of the appearance of the Grand Canyon's earliest known life—the stromatolites of the Bass Limestone.

Between September and early November of our imaginary year, sedimentary rocks formed on top of the worn-down, flat plains of the continents; these sedimentary strata are called the Grand Canyon Supergroup. Tectonic movements uplifted and created faults in the Supergroup, which later eroded almost completely away. The Tapeats Sandstone, earliest of the canyon's Paleozoic layers, covered the remnants of the Vishnu basement rocks and

the Supergroup a long time ago—525 million years. But on our one-year scale, this occurred on November 19.

The Grand Canyon's top layer, the Kaibab Formation, formed on December 9, while most of Arizona was near or below sea level. Dinosaur bones and a mile of sedimentary rocks accumulated starting December 13, but the dinosaurs and those rocks eroded away by December 26. Cutting of the Grand Canyon was largely completed by 2:00 p.m. on December 31. From 11:04:00 to 11:59:49 p.m. on December 31, lava

flows repeatedly dammed the Colorado River in the western Grand Canyon, forming lakes that lasted, on our scale, only a few seconds each. The earliest humans evolved in Africa on December 31 at about 8:08 p.m. but didn't make their way into the Americas until about 11:58:51. The ancestral Puebloans occupied Grand Canyon for five seconds from 11:59:50 to 11:59:55 on the 31st, just a few seconds before midnight. John Wesley Powell floated down the canyon one second before midnight, and Grand Canyon National Park was born about a half-second before midnight on the 31st.

The Grand Canyon Supergroup formed between 1,250 and 740 million years ago—or between September and early November when the geologic history of the Earth is compressed into one year.

and formed the Tonto Group between 525 and 505 million years ago. Then the sea retreated for more than 150 million years, spanning the entire Ordovician and Silurian periods. When the sea returned, it deposited the Temple Butte Formation, 385 million years ago. As a result, the Temple Butte's contact with the underlying Tonto Group is an unconformity.

The process is like making a two-layer cake, except that the top layer formed 120 million years after the bottom layer. The Grand Canyon's unconformities actually account for more time than the rock layers themselves record. Most of the Grand Canyon's time is invisible—that is, the time passed, but it left no geologic record.

When rock layers are missing, any fossils they may have otherwise contained are missing as well. Sometimes paleontologists can fill in the gaps by looking elsewhere—even on other continents— to find rocks from the same time periods as the Grand Canyon's missing ages. The unconformity between the Kaibab Formation at the top of the Grand Canyon and the human-laid blacktop path at the canyon's edge encompasses the entire Mesozoic and Cenozoic eras, a span of 250 million years. Yet similar sediments of the missing ages, and the fossils they contain, occur close by in Utah.

A portrait of the Swedish naturalist Karl Linne

The Linnaean Hierarchy

All organisms, whether they're alive today or now exist only as fossils, are classified according to the Linnaean Hierarchy. Karl Linne (sometimes called Linnaeus, 1707–1778) was a Swedish naturalist who developed this method in 1758.

Following Linne's rules, each organism gets a Latin species name that includes the genus—for example, *Homo sapiens*. The species name, *sapiens*, is never written without the genus, *Homo* (or an abbreviation of the genus, such as in *H. sapiens*). However, the genus (plural: genera), *Homo*, can stand alone. Both names are written in italics, with the genus capitalized and the species lower-case. Knowing that, you can discuss species—existing or extinct— with a Bolivian or a Turkish scientist, because the rules, and the Latin, are the same all over the world.

Each level of the hierarchy, called a taxon (plural: taxa), includes all the levels below it. A taxon is a group of classified

organisms, such as the Order Primates. All the families, genera, or species under the Order Primates, from people to monkeys, are also taxa. Here are the main classification levels of the hierarchy using the example of *Homo sapiens*, the human species:

Kingdom: Animalia
Phylum: Chordata
Class: Mammalia
Order: Primates
Family: Hominidae
Genus: *Homo*
Species: *sapiens*

Now that we've taken a quick look at how the fossils formed in the Grand Canyon and at what we call them when we find them, let's put on our hiking books and take a trip into the past.

Unconformities are gaps in the rock record due to erosion or nondeposition. This photograph shows the layered Tapeats Sandstone on top of the metamorphic Vishnu basement rocks—rock layers that formed about 1,200 million years apart. This gap—known as the Great Unconformity—is one of many unconformities in the Grand Canyon.

PRECAMBRIAN TIME IN THE GRAND CANYON

The Inner Gorge of the Grand Canyon features the Colorado River slicing through the Vishnu basement rocks—primarily schist, granite, and gneiss. These date back as far as 1.84 billion years.

The best way to understand fossil life in the Grand Canyon is to move forward in geologic time, period by period or era by era, and examine those creatures and plants that lived in each time. We'll start at the beginning, when the fossil record reveals the first stirrings of life in the area.

Vishnu Basement Rocks

The Grand Canyon's oldest rocks are the schists, gneisses, and granites of the Inner Gorge. These rocks are informally known as the "Vishnu basement rocks." Collectively, the Inner Gorge basement rocks are not only the oldest in the Grand Canyon but also in Arizona. Half of these, the schists and other metamorphic rocks, are about 1.7 billion years old. The other half are igneous intrusions called plutons—granite and gneiss, with radiometric dates ranging from 1.84 billion to 1.35 billion years ago.

Some of the Vishnu basement rocks of the Grand Canyon

The basement rocks are the rugged, dark-gray masses with the pink or white intrusions visible at the bottom of the canyon below Grand Canyon Village. Here the Vishnu basement rocks rise in jagged cliffs about 1,400 feet above the Colorado River.

The Vishnu basement rocks formed during a tectonic collision between volcanic islands and the continent. Before this event, which is called the Yavapai Orogeny, what is now Arizona was not yet part of the continent, but was a series of islands. Many such mountainous volcanic islands exist in the world today, such as Indonesia and the Aleutians. Most such islands will likewise accrete to their respective continents over time.

Warping and metamorphism of the Earth's crust normally accompanies tectonic collisions. As a result, any fossils the Inner Gorge rocks may have contained would have been obliterated by heat and pressure. Elsewhere in the world, relatively unaltered sedimentary rocks twice as old as the Grand Canyon's have yielded primitive microscopic fossils.

The Grand Canyon Supergroup

The Grand Canyon's oldest fossils appear in the sedimentary rocks, visible in some areas of the canyon, that overlie the Vishnu basement rocks. These rocks are known as the Grand Canyon Supergroup. Geologists date them from about 1.25 billion years ago to about 740 million years ago. These layers, visible from Grand Canyon Village and from eastern viewpoints such as Desert View, are

Stromatolites, made from cyanobacteria, leave behind fossils in a variety of shapes, including this layered example.

tilted about 15 degrees to the northeast. They appear in large, fault-bounded, down-dropped blocks, sandwiched between the Vishnu basement rocks below them and the flat-lying Paleozoic strata above. The Supergroup consists of the Unkar Group, the Nankoweap Formation, the Chuar Group, and the Sixtymile Formation, with a number of distinct strata in each group.

While the fossils in the Supergroup are primitive, they show remarkable diversity. The rocks reveal a variety of stromatolites and microfossils, including acritarchs. Stromatolites are limestone structures made by blue-green algae, properly called cyano-bacteria because they are photosynthesizing bacteria, not true algae. Stromatolite structures are visibly laminated, as each sunny day the algae formed another thin layer. Some stromatolites are spherical from rolling about in waves or currents. Others are domed, matted, or mushroom-shaped. Stromatolites likely covered a wide expanse of sea bottom during the Proterozoic. The Unkar Group contains many stromatolites in its lowest layer, the Bass Limestone.

More abundant and varied than the stromatolites of the Unkar Group are the acritarchs and other microfossils of the Chuar Group, which overlies the Unkar. The only place on Earth that the Chuar Group is found is in the far eastern part of the Grand Canyon, visible from Desert View. These are the tilted layers west of the Colorado River in the area called Chuar Valley.

Acritarchs are spherical cyst-like fossils found in ancient rocks worldwide. They appear to be the resting stages of unknown types of algae or other organisms. Grand Canyon Chuar specimens correlate with forms about 700 million to 800 million years old. One kind of Grand Canyon acritarch that has drawn special interest because of its large size is the spherical *Chuaria circularis*. It averages about a sixteenth of an inch in diameter—a giant among acritarchs.

In addition to stromatolites and acritarchs, one Chuar shale horizon contains curious vase-shaped microfossils in great abundance—10,000 fit in a cubic centimeter of rock. There are three types, each of which has an opening in the top of the "vase" with different ornamentation.

Microfossils from the Chuar Group also include tiny spheres, probably of cyanobacteria, and filaments described loosely as thallophytes, a term referring to primitive single-celled algae, fungi, or bacteria. Some of the spheres may also be Archaea, which are associated with modern-day hot springs, deep-ocean "black smokers," and other extreme environments. Archaea also occur in more "normal" environments. They were previously thought to be bacteria but now are known to be different, biochemically and genetically, from any other organism.

The Moon and the Mudflats

More than 1 billion years ago, the Earth revolved about 500 times a year—far more than the 365 revolutions we spin through annually today. Also, the moon was much closer to Earth. This combined to create enormous, fast-moving tides. The tides raced across coastal mudflats, ebbing and flowing many miles over a barren land.

Numerous tidal channels drained these flats as the ocean receded. Some tidal channels were broad and swift flowing, like rivers. However, they were not normal rivers, because they were the drainage of high-tide saltwater returning to the sea. The tidal channels had pebbly beds. The layered muds formed the Bass Limestone, the bottom layer of the Unkar Group. It is visible

The dark ledges below the surface of this ridge are layers of Bass Limestone.

A rock is an aggregate of mineral grains. The grains may be of any mineral; quartz, feldspar, clay, calcite, and dolomite are the most common in sedimentary rocks. Rock grains may be of any size, and the rock may have grains of one or more minerals.

As discussed earlier, the three major categories of rocks are igneous, metamorphic, and sedimentary. Igneous rock is any rock that melted and hardened again, metamorphic rock is any rock that has changed in texture or mineralogy as a result of heat and pressure (but not enough heat to melt it, or it would be igneous), and a sedimentary rock is formed by erosion of dissolved or granular sediment. The four-part process that forms sedimentary rocks begins when solid rocks weather and erode, forming sediment. The sediment is transported, and it is eventually deposited in a lowland or marine basin. The sediment then lithifies (hardens) to form a sedimentary rock. Most of the Grand Canyon's rocks are sedimentary, and sedimentary rocks are by far the most likely to contain fossils.

Sedimentary rocks may be clastic or nonclastic. Clastic sedimentary rocks are those that were transported in the form of individual grains. The transporting agent is usually water, wind, gravity, or ice. The grains, technically called clasts, may be of any size.

Here are three examples of clastic rocks.

Conglomerate: pebble to boulder-size grains

Sandstone: sand-size grains

Mudrocks: rocks made of mud, which is silt and clay

Of the clastic rocks, mudrocks are best for preserving fossils. There are four main types of mudrocks:

Siltstone: silt is like sand but the grains are too small to see with the naked eye

Claystone: a rock made from clay minerals

Mudstone: a massive, solid rock of silt and clay

Shale: a rock made of silt and clay that is flaky; it chips or splits easily into flakes or thin layers, and it erodes more easily than mudstone

Nonclastic rocks were transported as dissolved minerals, not as grains. For example, rivers can transport dissolved salt. When the salt precipitates and forms a rock (halite), that rock is nonclastic. Limestone, the most common nonclastic rock, is also transported in dissolved form, specifically as calcium and bicarbonate ions (atoms with an electrical charge). When the dissolved calcium and bicarbonate reach the ocean, shellfish extract these chemicals to make their shells. The shells are calcite and aragonite, which are the minerals that make up limestone. When a shellfish dies, the shell weakens, and waves and currents may grind the shell to bits. These bits of calcium carbonate can later harden to form limestone, an important fossil-bearing rock in the Grand Canyon and elsewhere.

Hermit Shale, a mudrock in Grand Canyon, preserves this fern.

as dark-brown ledges just above the schist of the Inner Gorge, far below Grand Canyon Village and in scattered locations elsewhere, including below Grandview Point and Desert View.

The pebbles in those ancient tidal channels were made by cyanobacteria, the earliest type of life found in Grand Canyon. The pebbles were pea-sized (pisolites), radish-sized (oncolites), or biscuit-sized (biscuits).

When the tide turned, the channels and pebbles were at rest. If it was daytime, the photosynthetic cyanobacteria grew long, sticky filaments. When the tide returned, it rolled the filament-covered pebbles. This caused them to trap particles of lime-mud sediment. In addition, the cyanobacteria's respiration created a surrounding aura that was depleted in carbon dioxide, and this caused additional limestone to precipitate from the seawater. The trapped and precipitated limestone particles formed a new layer around the pebble, like onion layers, as pisolites grew to become oncolites and then biscuits. Finally, the biscuits became too large to turn over, and they then tended to meld together in clumps.

While this was going on in the racing channels, the adjacent mudflats were alternately wetted by the tides and dried in the sun. When the tide was in and the muds were wet, algal mats grew, forming thin laminations—like layers of baklava pastries in the mud. Where shallow water drained through mud cracks, carving them deeper, the algal mat was divided into separate mounds. As the mounds grew higher and higher, they took on the mushroom shapes characteristic of some stromatolites.

In addition to these stromatolites, microscopic organisms abounded in the muds of the Grand Canyon Supergroup, especially in the Chuar Group. Of these, only *Chuaria* was visible to the naked eye. These blind, immobile bacteria appeared to be inactive, but within their tiny bodies they were metabolizing nutrients from the mud and reproducing prodigiously.

STRATIGRAPHY

Stratigraphy is the study of layered sedimentary rocks, or strata. Sediments are laid down in basins that may cover hundreds of square miles. Following the outlines of ancient seas, sediment layers grow progressively thinner toward the edges of their basins. Sedimentary layers are deposited horizontally, with the oldest layers at the bottom.

It is impossible to see a cross-sectional side view of strata unless, as at the Grand Canyon, the strata have been faulted or eroded. These and many other stratigraphic principles enable us to reconstruct ancient geographies and environments.

A formation consists of a single rock type or a sequence of related rock types. If one rock type—for example, sandstone—dominates in a formation, the layer is named accordingly: Coconino Sandstone. If there are several rock types, the word "formation" is used instead: Toroweap Formation. Formations are more dramatically displayed in the Grand Canyon than anywhere else in the world.

If a set of formations is inter-related in some way, such as by deposition in a single sea, then the set is called a group. Groups can be combined into supergroups, such as the Grand Canyon Supergroup. A formation may be subdivided into smaller units called members, and members may be further subdivided into beds.

THE CAMBRIAN AND DEVONIAN PERIODS

The Earth spun, the waters rose and fell, and animals died, their fossilized remains waiting for us to find them. Rolling on through geologic time, we come to periods that saw the birth of the first creatures with eyesight, the trilobites.

Fossils of the Cambrian Tonto Group

During the Earth's first 4 billion years, the continents grew by accretion and drifted about by plate tectonics. Icy mountains and burning deserts came and went unseen. Then, about 540 million years ago, a momentous event occurred. Trilobites became the first creatures (that we know of) to evolve eyes and actually see the world.

To see is to comprehend; we say, "I see" to express understanding. Not that seeing is the only way to understand, but our language equates the two—perhaps because sight so enhances comprehension. The trilobite took a giant leap in moving life along the evolutionary road we still travel.

The remarkable eyes of the trilobite, which are shown as the yellow-orange dots above, fascinate paleobiologists with their complexities. This trilobite was not found in Grand Canyon.

Even by modern standards, the trilobite had astonishing eyes. Many trilobite species evolved protruding crocodile-like eyes, faceted like those of insects. The trilobite kept a 360-degree vigil through myriad complex lenses—up to 15,000 per eye. Some of these fossilized eyes are sufficiently preserved for paleobiologists to dissect the trilobites' eye lenses under a microscope. In modern times, astronomers have rediscovered these lenses for use in telescopes: the Huygens and Descartes doublet eyepieces.

Trilobites are classified with crustaceans, insects, and spiders in the phylum Arthropoda, the jointed-footed animals. Of the arthropods, trilobite fossils are most abundant worldwide, especially in

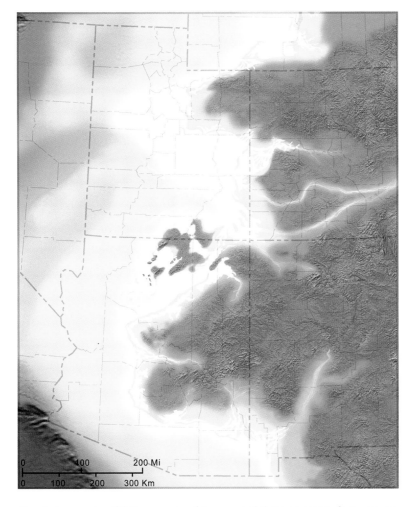

This map shows how inland seas
covered most of Arizona
during the Cambrian Period.

Cambrian rocks. Of thousands of known trilobite species, forty-seven
have been found in the Bright Angel Shale of the Grand Canyon.

With such diversity, trilobites evolved a complex ecology, fit-
ting into many niches. Some were free-swimming (nektonic); others
were bottom-dwelling (benthonic). Their feeding styles included
filter-feeding, scavenging, grazing, predation, and detritus-feeding
(burrowing in mud). Trilobites molted as they grew, shedding their
carapaces. They left their cast-off molts as fossils, so each trilobite
had the potential to leave several fossils of increasing sizes.

Grand Canyon trilobites are small, ranging in size from
less than half an inch to about four inches, in the case of *Anoria*.

However, there is tantalizing evidence of a much larger animal that may have preyed on the trilobite: *Anomalocaris*. One of the photos illustrated in Edwin McKee and Charles Resser's seminal work, *Cambrian History of the Grand Canyon Region* (1945), shows an elongated imprint that looks suspiciously like one of the feeding arms of *Anomalocaris*.

The trilobites lived in what is now known as the Tonto Group, which formed about 525 million to 505 million years ago. The three formations of the Tonto Group are, in ascending order, the Tapeats Sandstone, the Bright Angel Shale, and the Muav Limestone. Many additional creatures populated the Cambrian sea of the Tonto Group. Most abundant were the ones we know least about: the soft-bodied animals. They were squishy, and squishy doesn't fossilize well. Rather than fossilized bodies, we see fucoids, the burrows and trails those animals left in mud, now hardened into shale.

The Cambrian Period was a time of pervasive experimentation in new life-forms. Almost all of the basic modern groups of organisms, the phyla, first appear in the fossil record in the Cambrian. Representatives of these phyla evolved into all the animals and plants we know today.

Some scientists have suggested that terrestrial organisms might have existed even as early as the Precambrian Era, but little hard evidence has come to light. Exciting recent discoveries in the Bright Angel Shale of the Grand Canyon suggest the presence of primitive freshwater plants called bryophytes (mosses are an example). Until these discoveries, no confirmed Cambrian freshwater plants were known anywhere in the world. According to this evidence, freshwater streams carried microscopic "cryptospores" of terrestrial plants. The streams entered the shallows of the Tonto Sea, where the cryptospores sank into the bottom mud and remained as fossils.

Life Under the Tonto Sea

For the most part, the Tonto Sea was less than one hundred feet deep. The water was murky and the topography rugged. Ridges of hard Shinumo Quartzite rose from the bottom and emerged as islands above the waves. The pounding surf undermined sea cliffs and dashed boulders into the shallows, sometimes cascading in landslides

The echinodermata are the spiny-skinned invertabrates. The starfish is an echinoderm, but crinoids, blastoids, and sea urchins are the most common echinoderm fossils in Paleozoic rocks. All the echinoderms have a pentamerous (five-sided) symmetry, and most are spiny.

Class Crinoidea, the Sea Lily

Sea lily is a common name for crinoids because of their resemblance to flowers, although they are animals. Crinoids (and blastoids) have an appendage called a holdfast that functions as "roots," a column that is like a stem, and a crown that looks like the "flower" on top. The echinoderm skeleton tends to disarticulate (fall apart) after death, so fossil hunters usually find only pieces. The most common are sections of the column, called columnals or crinoid stems. Crinoid columnals are very common, but most of those at the Grand Canyon are very small. Many are less than an eighth of an inch in diameter.

Class Blastoidea, the Blastoids

Blastoids are extinct—unlike crinoids, which still live. Blastoids are similar to crinoids, but with a crown that has a nutlike shape and doesn't disarticulate after death.

Class Echinoidea, the Sea Urchins

Some echinoids resemble pincushions. Echinoid spines may be thin and sharp, or stubby and bumpy. Like most echinoderms, their skeletons tend to disarticulate after death, so only spines and skeletal plates are found for most species. However, sand dollars and burrowing sea urchins often stay intact after death.

A remarkable crinoid in the Redwall Limestone

into deeper water. One can still see these boulders above and west of Phantom Ranch, surrounded by and buried in Tapeats Sandstone.

At four inches, *Anoria* was the largest of the trilobites in the Tonto Sea. Along with the trilobite, the Tonto Sea was home to many diverse life-forms. There were tall eocrinoids, long-armed relatives of starfish that looked more like flowers than animals. Bivalved brachiopods lived in packed shoals with shells the size and texture of fingernails. Sponges were there too, mostly small and conical like thimbles but some larger, with calcareous skeletons. Jellyfish floated above waving colonies of algae.

The Enigmatic Temple Butte Formation

Perhaps Grand Canyon's least-known formation is the Temple Butte Formation, laid down during the Devonian Period. Sandwiched between the underlying Muav Limestone and the overlying Redwall Limestone, the similar-to-both Temple Butte Formation is diffi-cult to recognize. Most of it is purplish, has a sugary texture, and sometimes appears gnarled. It is easiest to recognize in channel-fill deposits in Marble Canyon along the Colorado River, but in most areas it is inaccessible, located partway up a sheer cliff. In much of the eastern Grand Canyon, the Temple Butte is missing altogether, but it becomes thicker as you head west. The prehistoric environ-ment of the Temple Butte was a shallow sea in the western Grand Canyon and a series of estuarine channels in the east.

PHYLUM PORIFERA: THE SPONGES

The sponge is so simple! No bones, no internal organs, no tissues—just a colony of cooperating single cells. Some sponge cells capture particles of food by using tiny tentacles called flagellae, which create a current that flows through the sponge. Still others wan-der amoeba-like through the sponge, digesting and distribut-ing food. Other cells secrete skeletal supports called spic-ules—little needles or stars of silica or calcium carbonate that give the sponge form. Because their cells are so loosely organized, sponges regener-ate easily—new ones can grow from broken parts. Because most sponges are soft, they are scarce in the fossil record.

The phylum Brachiopoda was immensely important in Paleozoic seas. While a few brachiopod species live today, clams have usurped most of their former niches. Like clams, brachiopods are bivalves, having two shells. The top and bottom shell are each called valves. However, brachiopod shells differ from clamshells. Most clam valves are mirror images of one another, like a pair of hands. Most brachiopod valves, however, are not alike. Often, one brachiopod valve has a ridge while the other has a groove or valley. Other brachiopods have one deeply curved shell and one flatter shell.

In combination with other fossils, brachiopods can indicate the relative age of a geologic formation. Species that lived for only a short time in only one formation or part of it are called index fossils because they uniquely identify that layer. Several of the Grand Canyon brachiopods are good index fossils. In the Grand Canyon, brachiopods occur in every oceanic formation from the Bright Angel Shale to the Kaibab Limestone.

After the Cambrian Muav Limestone was deposited, the area that would one day become Arizona was evidently above sea level for about 120 million years before the sea returned to deposit the Temple Butte. During this huge stretch of time, much longer than the entire Cenozoic Era, life continued to evolve. No record of this evolution exists at the Grand Canyon, however, because there are no rocks from the Ordovician or Silurian Periods (see the chart on page 12). But elsewhere in the world, the Devonian record shows a vast expansion of life: the first organic reefs, the first forests, the first amphibians, and an explosion in the fish population. In fact, the Devonian Period is informally known as the Age of Fishes.

Even more difficult to find than the Temple Butte Formation itself are any signs of life in its rocks. Other than stromatolites and conodont microfossils, the only other fossils found in this layer in the Grand Canyon have been two fish. One has not been specifically identified; the other is *Bothriolepis*, common in Devonian rocks worldwide. The little fish *Bothriolepis* was just four inches wide and a foot long. It is the most common placoderm fossil in the world. With its flat, armored head and thorax, its jointed crab-like pectoral fins, and its closely spaced eyes near its nostrils, *Bothriolepis* looks almost cute, in an alien sort of way—more like a giant water bug than a fish. *Bothriolepis* lived in the shallows and may even have inhabited freshwater. Some paleontologists think it had lungs and could crawl about on land using its articulated, armored fins. Others speculate that it used the arms to bury itself in the mud of drying

A stromatoporoid fossil from the Temple Butte Formation. Scientists are uncertain why so few fossils have been found in this Grand Canyon rock layer, and why many that are found there are worn and fragmented.

ponds, waiting in torpor for the next rain, as many small arid-land animals do in times of drought.

Elsewhere, Devonian strata contain a few primitive jawless fishes (the agnaths), along with sharks, ray-finned and lobe-finned fishes, and curious armor-plated fishes called placoderms, whose oddest examples are the antiarchs, such as the Grand Canyon's *Bothriolepis*. Near Flagstaff, Arizona, fifty miles from Grand Canyon when not relying on roads, about ten types of fishes, including *Bothriolepis*, were found in a small Temple Butte deposit. In the Temple Butte Formation west of the Grand Canyon occur scant marine fossils—corals, gastropods, and brachiopods, worn and fragmented almost beyond recognition.

Scientists have not determined why so few fossils have been found in the Temple Butte, while fossils are abundant in rocks of the same age in central and southern Arizona. One reason may be the inaccessibility of Temple Butte cliffs, preventing easy searching. Another reason is that the formation is mostly dolomite. Dolomite is a rock that was originally limestone (calcium carbonate), but magnesium-rich seawater filtered through the layers, changing the limestone to dolomite (calcium-magnesium carbonate). The process of changing limestone to dolomite can destroy fossils. It seems likely, however, that the reasons for this fossil rarity originated in the ancient environmental conditions. For example, perhaps shells were pulverized by wave action, and the dolomite was originally a sandy beach or mud made of shell bits, later blurred by dolomitization.

Devonian Mass Extinction

The extinction of the dinosaurs 65 million years ago is the best-known example of mass extinction—the extinction of many types of organisms at once. But it was not the only one. A worldwide mass extinction occurred soon after the Devonian Temple Butte Formation was deposited. In the Devonian mass extinction, land

This painting shows what a massive meteor strike might have looked like. Some scientists point to such strikes as a cause of the Devonian mass extinction.

The Cnidaria include jellyfish, hydroids, sea anemones, and corals. Of these, only corals secrete a hard skeleton (made of calcium carbonate) and are common in the fossil record.

The Cnidaria took an evolutionary step beyond the simple sponges, developing discrete tissues for different functions. They have a ring of tentacles that contain stinging nematocysts for capturing prey that wanders too close. They have a "body bag" that includes epidermis and endodermis. They also have a primitive gut.

Some Paleozoic corals (Rugosa) were solitary and others were colonial (Tabulata). Both left an extensive fossil record.

A fossilized coral. The rust-colored areas in the photograph are lichen growing on the rock surface.

life was not much affected. But the nearshore marine life was decimated, including reef-building stromatoporoid sponges, primitive coral taxa, trilobites, conodonts, and several kinds of fish. No clear evidence explains how the Devonian extinction came about. It has been tentatively attributed to a worldwide cooling that accompanied an ice age in the southern continent of Gondwanaland, one of two huge land masses that later separated into the continents we know today. Other scientists suggest a meteorite may have contributed to the Devonian extinction.

How might an ice age affect sea life but not land life? If the ice were confined mainly to the poles, land life in temperate areas might survive the ensuing cool spell. But when oceans freeze at the poles, water is locked in ice and the sea level drops. During much of the Devonian Period, extensive inland seas teemed with life. But as the sea level dropped and inland seas retreated, reefs were left high and dry. Not just the reef-building animals and plants were affected. The drying of the reefs would have affected the fish along with the myriad other organisms dependent on that habitat. So, for the Devonian mass extinction, an ice age, rather than a meteorite, seems the most likely cause.

Bothriolepis and the Devonian Ice Age

Mentioned earlier, one of the most intriguing creatures of the Devonian Ice Age was the antiarch placoderm *Bothriolepis*. While its predators, such as bigger placoderms and sharks, could not survive in brackish water, *Bothriolepis* adapted to fresher water and maybe even to land—it had lung-like organ cavities hidden in its armor casing. Scientists believe that at some point *Bothriolepis* took advantage of its multiple capabilities and moved across the land into safer waters. Leaving one body of water behind, the creatures used their arms to pull across wide mudflats until they came to ponds. There they buried themselves in mud to await a safer time, usually the warmer climate of spring. This ability to live in and out of the water helped *Bothriolepis* survive amid a changing and increasingly difficult environment.

A model of an artist's conception of *Bothriolepis*, an animal that might have been able to live on land and in the sea.

THE LATE PALEOZOIC ERA

The massive cliffs of the Redwall Limestone still hold fossil secrets from the past.

The Redwall and Surprise Canyon limestones contain many fossils. In the next set of layers—the Supai, Hermit, and Coconino—the fossil record is not as clear as it was in the underlying limestone layers. Still, a little bit of ingenuity and inference can reveal what life was like for the creatures of each time period.

The Redwall Limestone and the Age of Crinoids

The Redwall Limestone in the Grand Canyon forms a massive red cliff that in some places, in combination with the Muav and Temple Butte Limestones, exceeds 1,500 feet in height. You can see the cliff from almost anywhere, about halfway down the canyon walls. Because the Redwall is an inaccessible, sheer cliff in most areas, its fossils are best hunted where it is exposed at river level along the Colorado River or in boulders of Redwall that have tumbled down onto the Tonto Platform below. The limestone itself is gray, but its cliffs appear red in most areas because they are stained by runoff from the iron-bearing Supai and Hermit formations above the Redwall. Many alcoves and caverns penetrate the cliffs. Here and there, freshwater springs plunge from Redwall caves.

Inland seas can became heavily salinated or stagnant (oxygen-deprived) because of limited circulation (having fewer or shallower outlets to the open ocean). However, the vast seas of the Mississippian Period that formed the Redwall and its limestones contained clear, normally saline water and carried in their currents a high level of nutritious organic material, food for the myriad filter-feeding organisms. For roughly 20 million years (350 million to 330 million years ago), those seas extended into most of Arizona and far inland across North America. They fronted on open ocean to the north, south, and west.

The Mississippian Period was a heyday for sea creatures. The entire seafloor became a zoological garden. While crinoids were most common, the sea also supported dense populations of brachiopods, corals, bryozoans, gastropods, echinoderms, worms, conodonts, sponges, sea cucumbers, anemones, clams, sharks, and bony fish. Life was so abundant that thick layers of living, shelled organisms built up over large graveyards of invertebrate skeletons, called biostromes. Other skeletal remains piled into extensive, high mounds called bioherms or patch reefs. Of all this life, the most common types

This map shows the Redwall sea covering nearly all of what is now Arizona and the Southwest. As it receded, it left behind a wealth of fossilized creatures.

The conodont animal is a tiny chordate. Its mouth parts are abundant in Paleozoic rocks, although, rarely, other parts of the animal are found. The mouth parts never show wear, so they must have been jaws that snatch rather than munch, and their prey was probably soft-bodied. Paleontologists believe conodonts were small eel-like animals that lived buried in mud.

Conodonts evolved rapidly, so they are useful stratigraphic indicators. Each species existed for only a short time, so that when we find one, we know rather precisely the age of its enclosing sediments. Paleontologists working for oil companies use conodonts to correlate rocks from Texas to Saudi Arabia and beyond. There is an entire paleontologic specialty called conodont biostratigraphy, and because it helps locate oil, it is one of the highest-paying fields in paleontology.

preserved as fossils are the corals, gastropods, brachiopods, and, especially, crinoids. In most of the Redwall, the limestone is made almost entirely from crumbled crinoid skeletons.

A Redwall Disaster

Scientists believe more than a million species lived on the reef, yet only a minuscule few left behind fossils. In modern reef environments, most organisms were attached to the bottom, and most fed by filtering organic material from the seawater. *Cladodus* was a small shark that swam the reef-metropolis. *Cladodus* preyed on the tiny bony fish that made their homes in the numerous niches of the reef environment.

Two types of predators were of concern to *Cladodus*. One was the larger sharks, whose niches were mainly in the open sea above; the other was the three-foot-long shelled nautiloid *Rayonnoceras*, whose habitat was mostly to the east in shallower water. To avoid these nautiloids, *Cladodus* stayed farther out to sea.

In this water world, *Cladodus* undulated over the seafloor above clusters of rugose (wrinkly) corals shaped like six-inch cones and topped by brightly colored polyps with stinging tentacles. *Cladodus* passed masses of the colonial tabulate coral *Syringopora*, covered by much smaller polyps. Wedged in gaps and crevasses, anemones waved their tentacles, trapping fish and snails but harboring tiny fish immune to their stings.

There were forests of living crinoids, looking like colorful margarita parasols. Their stems varied from a few inches to several feet tall. Millions of years later, when the Redwall Limestone solidified, most of its 500-foot-thick layer would be composed of the remains of these crinoids.

Partway through the deposition of the Redwall Limestone, an enigmatic event occurred. Geologists have long recognized a limestone bed strewn with *Rayonnoceras* shells. Remnants of this nautiloid graveyard can be found from Grand Canyon to Las Vegas, Nevada. The cause of this mass mortality is unknown.

The Mississippian Surprise Canyon Formation

Approximately 340 million years ago, the shelf that formed the Redwall retreated, leaving its limestone several hundred feet above sea level and exposed to erosion. The limestone was soluble in rain—and groundwater. Caverns opened in the earth, and a karst landscape slowly evolved. Karst is a landscape characterized by rock dissolution and its resulting topography—caves, sinkholes, steep-sided canyons, potholes, and underground lakes and drainage systems.

The Surprise Canyon Formation, an outcropping of which is seen as a wedge-shaped cliff in the center of this photograph, is one of the most fossil-laden rock layers in the national park.

When the sea returned to the west of the Grand Canyon in the late Mississippian Period, about 320 million years ago, it created the Surprise Canyon Formation, which was first recognized as a separate formation in 1979 and formally described in 1985. Streams flowing west to the sea formed the channels containing the Surprise Canyon Formation.

As the Surprise Canyon sea encroached, it deposited sediments in these stream channels and also in caverns in the Redwall karst. Western ends of the stream channels became marine estuaries, as shown by their fossils. In the western Grand Canyon the channels were up to four hundred feet deep and half a mile wide, but in the east, in the village area, most channels are less than fifty feet deep. Visible cross sections of the channels have flat tops and rounded bottoms.

Surprise Canyon sediment deposits have yielded one of the richest fossil troves in Grand Canyon. More than sixty species of fossils occur in the formation. They include brachiopods, gastropods, clams, corals, bryozoans, crinoids, blastoids, stromatolites, trilobites, bone fragments, shark teeth, twelve species of terrestrial plants, and many microfossils (ostracods, foraminifera, and twenty-two species of spores).

More than sixty species of fossils have been found in the Surprise Canyon Formation, including this coral fossil, *Michelinia* sp.

Except for estuarine incursions, the shoreline remained west of the Grand Canyon. At the end of Surprise Canyon time, the shoreline retreated farther westward. The landscape became increasingly terrestrial, with deltas, deserts, and floodplains—now represented by the Supai Group, Hermit Formation, and Coconino Sandstone.

Fossils of the Red Beds: Supai and Hermit

Five rust-red formations accumulated at the Grand Canyon during part of the Pennsylvanian and Permian periods, from about 315 million to 280 million years ago. Geologists have combined the bottom four of these into the Supai Group. The fifth formation, on top of the Supai, is the even redder Hermit Formation. In the Grand Canyon, Supai strata erode into alternating ledges and slopes, while the less durable Hermit Formation tends to form slopes. The entire sequence of rock layers consists of cross-bedded sandstones, siltstones, mudrocks, and combinations of these, with a few marine limestone beds in the Supai. The lower three Supai formations are of Pennsylvanian age, while the uppermost formation of the Supai Group (the Esplanade Sandstone) and the Hermit Formation are Permian.

Recent studies show that desert winds formed most of the Supai red beds. The deserts lay on a broad coastal plain on which lazy rivers meandered. Shifting seas occasionally inundated the plain, leaving behind limestone beds that contain marine fossils such as brachiopods. By Hermit time, the sea had withdrawn. The Hermit Formation is considered terrestrial, a widespread fluvial (river and stream) environment containing many land-plant fossils but no marine limestones with marine fossils.

By the Permian Period, flying insects included the dragonfly *Typus whitei*, with an eight-inch wingspan. The Permian air may have contained more oxygen than today—perhaps 35 percent compared to today's 21 percent.

The erosion of the Hermit Shale has revealed thousands of fossils, especially primitive plants such as seed ferns.

This could explain why insects grew much larger than they do now. Insects lack lungs but take in oxygen through tiny pores in their exoskeletons. At 35 percent, oxygen would penetrate more effectively, allowing larger body size.

While there are no bone fossils in the Supai and Hermit mudrocks, sparse amphibian trackways have come to light. The most common red-bed fossils, however, are in the Hermit Formation:

fronds and stems of seed ferns (pteridospermophytes), joint-stemmed relatives of horsetails (sphenopsids), pines (coniferophytes), and ginkgos (ginkgophytes). Surprisingly absent from this list is any evidence of a true fern fossil. The semiarid climate of the Hermit Formation may have discouraged their growth.

In rainy climates outside the Grand Canyon area during the Permian Period, trees grew tall. Scaly barked lycopsids, jointed sphenopsids, and forty-foot fern trees abounded.

But in the Hermit landscape, drought stunted the xerophytic (drought-adapted) seed ferns and diminutive pines, which were not much taller than the local amphibian fauna. Fossilized plant material has not survived to the present. The plants left only their death-mask imprints in the mud.

The Coconino Sandstone: Fossil Tracks on a Sand Sea

The Coconino Sandstone may be the most obvious cliff in the Grand Canyon. It is the light-colored layer a relatively short distance below the canyon rim. The Coconino is entirely quartz sand—fine-grained, well-sorted, and well-rounded, thanks to endless years of blowing desert winds. In the Permian Period, about 275 million years ago, the Coconino desert stretched from Arizona to Canada.

The Coconino desert was an erg, a sand sea, perhaps like the modern Sahara or Namib deserts in Africa. Erg sand forms into all manner of dunes, but the largest and most characteristic of these are huge parallel sand ridges called draas, separated by interdune valleys. As the dunes migrate downwind, their spilling sands create sweeping laminations called cross-beds. These are oriented according to the prevailing winds, and in the case of the Coconino desert, the wind blew from north to south.

Across the Coconino dunes, a variety of vertebrate and invertebrate creatures walked and crawled. None, however, left bones or body fossils, only tracks and trails in wide variety. Ichnologists have

Ancient tracks, known as trace fossils or ichnocoenoses, are common in the canyon's Coconino Sandstone layer.

A woman examines amphibian tracks in the Coconino Sandstone seen along Soap Creek, a tributary of the Colorado River in the far northeastern corner of the national park.

named three species of the genus *Chelichnus* (formerly *Laoporus*), a presumed mammal-like reptile, from these tracks. There are tracks of a fourth species, probably unrelated, that are still pending classification. These mammal-like reptiles are tentatively classified with the plant-eating caseid family, a family that includes the fin-backed *Edaphosaurus*. All members of that family share the characteristics of a wide body and a sprawling posture. The Coconino species' tracks vary from shorter than an inch to longer than four inches, representing animals from the size of a house cat to that of a small alligator. The mammal-like reptiles of the Permian and Triassic periods would give rise to all modern mammals, including, in time, people.

Invertebrates that left traces in the Coconino Sandstone include scorpions, spiders, isopods, and millipedes. A number of scientists have reproduced the tracks of these invertebrates by having similar living creatures run uphill and downhill across sandy surfaces, some wet, some dry. Paleontologists have photographed the tracks and compared them with ancient trackways in the Coconino. By this method, they have been able to show which direction an ancient scorpion traveled.

It is interesting to consider how animals of many sizes could exist in such a dry desert. Where did they get food? What is the base of the paleoecosystem? Who are the primary producers of energy? Again, we can make analogies with modern sand seas. Most of the Namib and Sahara deserts receive less than an inch of rain in an average year. However, near the coast of the Namib, frequent fogs moisten the sand briefly in the mornings. When the precious moisture comes, life proliferates. Modern desert plants have many strategies for surviving in a dry climate. These include growing deep taproots, storing water in the trunk or roots, having a shallow but pervasive root system that reacts quickly after a rain, or living most of the time as seeds and sprouting only after a rain. Ancient desert plants would probably have used similar strategies.

Desert animals also have their drought-adaptive strategies. Many do not have to drink, deriving their water from moist plant foliage. Many are nocturnal or crepuscular, emerging from their burrows while the weather is cooler and the humidity higher.

In the Permian Period, flowering plants had not yet developed, but vegetation and algae must have taken advantage of the rare rains of the Coconino desert and briefly covered the dunes with an evanescent greenery. Lichens undoubtedly inhabited interdune flats. Some modern lichens can even detach from the ground and live a nomadic life, letting the wind blow them from place to place. Where the water table came to the surface, hardy plants would have taken permanent root beside oases and along ephemeral streams.

Vegetation was sparse, as were larger herbivores like *Chelichnus*, which left its tracks on Coconino dune faces. But looking closer, one might see an entire miniature ecosystem. Small invertebrates preyed upon one another or ate tiny plants. Small vertebrates may have lived either by preying on the invertebrates or by eating the plants that grew in the meager habitat. Larger

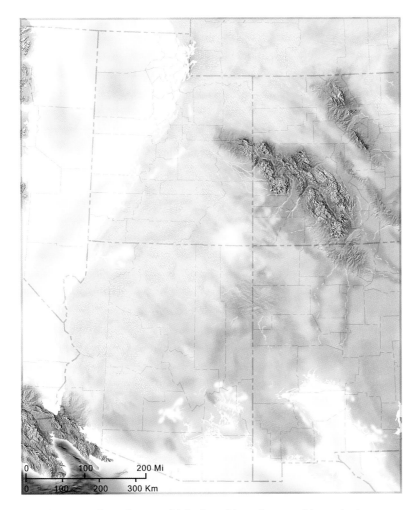

By the time of the Permian Period, as this map depicts, the ocean water had almost completely retreated from what is now Arizona, leaving vast tracts of land that became habitat for creatures such as mammal-like reptiles.

carnivores, when they could find nothing else, could comb the beach for stranded fish.

All we have today are the varied trackways in the Coconino Sandstone. But their makers could not have lived in a vacuum. They certainly occupied niches in an ecological system that included plants at the base of the food chain and larger carnivores at the top.

THE SEAS RETURN

The uppermost layers of the Grand Canyon are the Kaibab and Toroweap formations. But this was not always so. During the Mesozoic Era, long after the Kaibab and Toroweap formed, deserts and inland seas deposited sediments more than a mile thick on top of the current Grand Canyon strata. These sediments eroded away almost completely in what is now the Grand Canyon, leaving the scene as we see it today. The missing layers can still be seen to the north and east of the Grand Canyon.

Fossils of the Toroweap and Kaibab Formations

After an absence of about 20 million years, the ocean twice encroached from the west and deposited in its seabed the top two layers of the Grand Canyon. First came the Toroweap Formation, about 273 million years ago, then the Kaibab Formation, about 270 million years ago—both of the Permian Period. The Toroweap sea made only a shallow incursion, leaving mostly intertidal sediments—sand, silt, clay, and salts (gypsum and halite)—with just one deeper-water limestone layer. The Toroweap is the pink-and-tan vegetated slope just below the Kaibab wall at the canyon rim. In contrast, the Kaibab sea was deeper and filled with greater diversity and volume of life than the Toroweap, leaving sandy, cherty, cliff-forming limestones that contain numerous fossils.

The Toroweap and the Kaibab both show cyclic bedding. This is most obvious on the exposed Kaibab cliff that forms the canyon's rim. Here one sees a series of beds repeated many times as the sea became alternately deeper and then shallower, perhaps due to cyclic glacial buildup at the poles (two worldwide glacial periods occurred during the Permian). The cliff shows an imperfect pattern of prominent, rounded, gray ledges underlain by receding, rough, brownish layers. Each of the prominent gray layers formed in offshore subtidal areas with deeper water, and each contains numerous and diverse fossils. Because offshore subtidal areas were perpetually water-covered and tranquil, far below the energetic surface waves, they were often rich in life-forms.

The brown receding layers of the Kaibab Formation are sandy, silty, and mixed with chert. These layers are generally less fossiliferous, because they formed in shallower water in the tidal or subtidal zones. Here at low tide, the bottom-dwelling organisms were exposed to

In this photograph, the upper cliff layer is the Kaibab Formation and the sloping middle layer is the Toroweap Formation. Both layers formed from separate incursions of the sea.

Bryozoans almost always form small colonies of various shapes and most live in the sea. At first sight, the tiny (1/32 of an inch) polyp-like animals appear to be corals. Bryozoans are related to brachiopods, but look quite different. Their fossils are abundant. Bryozoans are common on modern seashores but are easy to overlook or to mistake for corals or seaweed. The animals differ from corals in that they have organ systems, including a complete digestive system.

Bryozoans also have muscles for retracting into their tiny chambers, tentacles for catching food particles, and reproductive systems. One can often recognize bryozoans by the tiny dots that cover the surface of a colony. Each dot is a zooecium.

waves or even to air, and this made life more difficult. Organisms best adapted to this high-energy environment were those with sturdy shells, especially clams, shelled cephalopods, and snails. The salt flats of the Toroweap Formation were even more hostile to living things. Except for within its one limestone layer, the Toroweap contains very few fossils.

Not a plant, but a colony of tiny animals, bryozoans such as this one flourished in the rich environment of the Kaibab sea.

Niches in the Kaibab Sea

While fossils are absent in much of the Kaibab and Toroweap, they are diverse and plentiful where they do occur. Counting both formations, the species list includes seven vertebrates, nine corals, ten crustaceans, thirteen cephalopods, sixteen bryozoans, thirty-four snails, thirty-four brachiopods, thirty-five clams, and one each of sponges, annelid worms, sea urchins, and crinoids. However, this fauna is less diverse than assemblages in Texas and elsewhere from the same period.

Each of these species occupied its preferred habitat and niche. The habitat is simply the place where an organism lives. The niche is the organism's function in the habitat: what it does, what it consumes, what it contributes, and how it interacts with other organisms and the environment. The niche has been likened to an organism's "job" in the habitat.

For example, all the following features constitute the niche of one type of Kaibab bryozoan, a tiny polyp-like creature that secreted a limestone colony. Its habitat was a muddy substrate populated by shellfish. The bryozoans multiplied to create a colony of hundreds or even millions of individuals. Each individual bryozoan created a tiny hollow chamber called a zooecium in the colony. The colony became a substrate feature up to a foot tall and wide. It provided habitat for a variety of unrelated organisms, including algae, brachiopods, and invasive microorganisms.

A brachiopod *Meekella* fossil in the Kaibab Formation, a species sometimes seen with bryozoans.

Such bryozoan colonies were important reef-formers in waters more tropical than those of the Kaibab and Toroweap. Here, the colonies were smaller but they continued to be important even after the death of the colony. Other organisms could adopt an abandoned colony and use it as a home for their own colony.

Another aspect of the bryozoan's niche was its method of filter-feeding. The bryozoan extended its lophophore, a tentacled, circular organ, into the open sea to extract particles of food from the current. The particles could be anything organic, including individual algal cells or tiny protozoa such as radiolarians. The bryozoan also provided food to other organisms. A number of small predators could attack and eat individual bryozoans. Certain fish and invertebrates could feed on the colony by grazing or even biting chunks. In reproducing, some bryozoans released sperm and eggs directly into the current. Others harbored the eggs in a brood chamber, releasing them when they grew into larvae. These immature organisms provided food for many other creatures.

Waste products and dead bryozoans' soft tissues constituted a final part of the niche. The bryozoans ejected their wastes directly into the sea, adding fertilizer to the substrate and food for decomposers such as bacteria. After death, the bryozoans' tiny corpses provided further nourishment for the decomposers. In similar fashion, each species in the Kaibab and Toroweap had its own habitat and niche, giving and taking in the bountiful sea.

Life in the Kaibab Sea

The amazing life that lived around and beneath the waters of the Kaibab sea is revealed through the fossil record—and a little imagination. At the shoreline and in the nearshore waters, one finds evidence of heavy surf action. Fossils are often broken and fragmented. In the sandy bottom, nautiloid cephalopods made their homes amid the churning surf. The Kaibab was home to many more brachiopods than one would find in modern seas. This reflects the Paleozoic abundance of brachiopods, which are only a minor feature of modern marine fauna. Empty brachiopod shells that we find are usually closed and together, while the two valves of clamshells are separate, having fallen apart. This is a result of the different configurations of ligaments and muscles connecting the shells. In a clam,

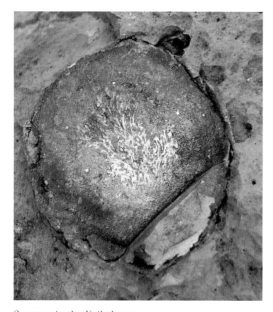

Sponges in the Kaibab sea eventually hardened into silica blobs forming what are today chert nodules.

the ligaments are like rubber bands that open the shell. The clam uses its muscles to hold the shell closed. When the clam dies and its muscles relax, the shell opens. In brachiopods, the rubber-band ligaments hold the shells together—its muscles are designed to open the shell as needed, so in death the ligaments keep the shell closed.

Farther out into what was the Kaibab sea—and is now the site of ranger-led fossil walks at Grand Canyon National Park—the most conspicuous life-form was the sponge *Actinocoelia*. Each sponge was a few inches wide and tall. Like many sponges, *Actinocoelia*'s skeleton was made of interlocking needles of silica called spicules, almost microscopic in size. When the sponge died, the spicules fell onto the ocean floor and attracted additional dissolved silica from the seawater. In time, each dead sponge became the nucleus of a silica blob, which then hardened into a chert nodule. Chert is the hardest sedimentary rock, and it eventually strengthened the Kaibab Formation, retarding erosion.

Fishes were not abundant here, but there were a few varieties, such as a strangely shaped petalodont shark with flat teeth.

The clam *Aviculopecten* was plentiful and large, up to six inches wide, the biggest shellfish in the area. Brachiopods were

attached to whatever hard substrate they could find, fastening themselves with spines and fleshy pedicles. The crinoids were small but plentiful, and snails glided slowly along. Evidence has also been found of bryozoans, jellyfish-like ctenophorans, and trilobites. Some of the brachiopod shells are found in various orientations. One reason for this is the interesting process of bioturbation, in which animals moving through an area change the position of fossils, rocks, and other sediments. This process can sometimes confuse the stratification of remains, posing a problem for paleontologists.

The enormous diversity of life highlights the fact that paleontologists are able to find only a small fraction of the organisms and species that lived in an area.

What was once the Kaibab sea is now a great place for visitors to take a ranger-led fossil walk along the South Rim of the Grand Canyon.

The phylum Mollusca includes clams, snails, cephalopods, tusk shells, and other, more minor groups. The minor groups are uncommon as fossils. But these four main groups, each its own taxonomic class, are common throughout the fossil record.

Class Bivalvia, the Clams

The clams (below) have two valves (shells) that are usually mirror images of each other, like left and right hands. A line drawn from top to bottom of a clamshell does not divide a single shell into mirror images, as it does a brachiopod shell. Again using the hand analogy, if you draw a line down a single human hand from top to bottom across the palm, the two halves would not be mirror images, because the thumb would be on one side but not

the other. Similarly, clams are different from brachiopods. So clam shells are equivalved but not equilateral, while brachiopod valves are the opposite: equilateral but not equivalved.

While clams are much more common than brachiopods in modern oceans, brachiopods were more common in Paleozoic seas.

Class Cephalopoda, the Nautiloids and Ammonoids

The cephalopods are the most advanced of all the invertebrates. The most familiar modern cephalopods are the octopus and squid, but these are soft-bodied and don't fossilize well. The ammonoids (now extinct, above) and the nautiloids have shells, so they are common in the fossil record. These marine creatures are essentially an octopus or a squid in a shell.

In contrast to the snails, cephalopod shells have many hollow chambers separated by bulkheads called septa. The animal lives in only the outermost chamber. A fleshy siphuncle connects all the interior chambers, enabling the animal to regulate its buoyancy and its upright orientation

by pumping fluid or nitrogen in or out. Among their highly advanced characteristics, cephalopods have excellent eyes, similar to the human eye. The largest eye in the present-day animal kingdom is that of the modern giant squid, whose eyes are the diameter of a dinner plate.

Cephalopods are jet propelled—they use a tubular siphon to shoot out water, which propels them quite rapidly. The siphon is muscled and can be aimed precisely and turned quickly to change direction.

Class Gastropoda, the Snails

The earliest true snails showed up in the late Cambrian Period. Unlike the cephalopod, the snail occupies its entire shell. Most gastropods (below) have a spiral shell that tapers to a point. Most also have an operculum that functions as a "door," closing the shell and serving as protection when the snail withdraws its body into the shell.

Snails have an advanced organ system. It includes a muscular foot equipped with mucous glands, a complete digestive system, and nervous and reproductive systems. Some snails are either male or female, others are hermaphroditic (both sexes in a single snail). Gastropods have a well-developed head, with eyes, mouth, and odor-sensitive tentacles that look like antennae.

Class Scaphopoda, the Tusk Shells

Scaphopod shells are not well known except to the shell enthusiast, but are fairly common as fossils (the fossil above is similar to ones found in the Grand Canyon). At the Grand Canyon they can be seen in the Kaibab Formation. Tusk shells have an opening at each end of the shell, the "head" emerging from the lower, larger opening. Scaphopods spend their lives partly or wholly buried in sand or mud.

THE MESOZOIC AND CENOZOIC ERAS

The fossil record of Grand Canyon begins with cyanobacteria and primitive plants and moves on to invertebrates. It closes with evidence of vertebrates: mammals that roamed what is now the American Southwest.

The Missing Rocks of the Mesozoic and Cenozoic Eras

To the east of Grand Canyon National Park, the Mesozoic Chinle Formation in the spectacular Painted Desert includes many fossils and traces of dinosaurs. Such layers are almost entirely absent from Grand Canyon, as they eroded away during uplift of the Colorado Plateau.

The Mesozoic Era is popularly known as the Age of Reptiles. Following the Mesozoic, the Cenozoic Era comprises most of the Age of Mammals. In Grand Canyon National Park, few Mesozoic or Cenozoic rocks or fossils exist because water, wind, and gravity have eroded them away. During a span of 251 million years, Mesozoic and Cenozoic sediments and rocks accumulated to more than a mile deep atop the Kaibab Formation. But uplift results in erosion, and with the rise of the Grand Canyon's high plateaus in late Mesozoic and early Cenozoic times, almost all these sediments and rocks were stripped from the area.

Mesozoic rocks still remain in abundance elsewhere on the Colorado Plateau in northern Arizona, much of Utah, southwestern Colorado, and northwestern New Mexico. Cenozoic rocks remain in the western Grand Canyon, but they are sparse and confined mainly to small river channels, lava flows, and lava dams. Geologists have found freshwater mollusks here. Of the Mesozoic and Cenozoic rocks that once existed in or near the eastern Grand Canyon, only two relatively small Mesozoic remnants remain: Red Butte near the South Entrance to the national park and Cedar Mountain near the Desert View Entrance.

While no fossils from rocks of these eras exist in Grand Canyon National Park, there is little doubt that many dinosaurs and mammals visited or lived in the Grand Canyon area long before the Colorado River cut the canyon. Christa Sadler's excellent book, *Life in Stone:*

Fossils of the Colorado Plateau (2005), gives many details about Colorado Plateau rocks and the flora and fauna they contained.

Quaternary Fauna and Flora in Grand Canyon

Late in the Cenozoic Era, sometime between 5.5 and 4.3 million years ago, the Colorado River established its current course through the Grand Canyon. Of course, the Grand Canyon was not excavated at a particular time but during a time span that started before 5.5 million years ago and continues today.

As the Colorado River excavated down to the Redwall Limestone, caverns formed, their entrances facing the wide Tonto Platform. A few of the caves were accessible to animals that inhabited the area during and after the Pleistocene ice ages, the last of which ended about 11,000 years ago. In these caves we find the remains of the extinct Shasta ground sloth (*Nothrotheriops shastensis*) and Harrington's mountain goat (*Oreamnos harringtoni*), as well as extant species such as the coyote (*Canis latrans*), the bighorn sheep (*Ovis canadensis*), the California condor (*Gymnogyps californianus*), the chuckwalla (*Sauromalus ater*), and the pack rat (*Neotoma* spp.), among others. In the caverns, we also find evidence of Native Americans from the Late Archaic culture. These

A cave, perhaps once home to animals or humans, looks out onto the Tonto Platform.

people entered the caves 3,000 to 4,000 years ago and left behind their bones and tools, as well as split-twig figurines. Because those fossils and artifacts are precious resources, access to the caves is restricted to scientific work authorized by the National Park Service.

The circumstances in which the fossils of the Quaternary Period (1.8 million years ago to the present) formed differ dramatically from those of the much older, rock-bound fossils. The rock-bound fossils left their remains locked in ancient Precambrian and

Paleozoic rocks, about 1.2 billion to 250 million years before the Colorado River cut the Grand Canyon. But the Quaternary fossils, found mostly in caverns, represent fauna and flora that lived within the Grand Canyon more or less as we know it today. The animals left their organic remains and dung, unpetrified, in the caves and rock shelters they occupied, not embedded in solid rock.

The Quaternary flora is no less interesting than the wildlife. Flora is an indicator of past climate. During the ice ages, which encompassed 90 percent of the past 1.8 million years, the Grand Canyon's climate was cooler and there was no late summer monsoon (rainy) season, as there is now. An open forest of juniper and pinyon pine occupied the Tonto Platform. Douglas-fir, limber pine, and spruce grew on sloping formations well below the South Rim.

Much of our knowledge of this Quaternary flora comes from pack rat middens. Pack rat middens resemble hard, dark chocolate

Pack rat middens in caves such as this one in Joshua Tree National Park in California provide a rich trove of information about ancient plants and climate.

cakes, although they don't smell as good. The middens are found in rock crevasses where generations of pack rats have lived. Middens are the front porches to the pack rat nest, made from discarded plant parts and feces, cemented by urine from pack rats that may have lived in the crevasse over hundreds of generations. By carbon-dating the midden from bottom to top, scientists can learn what plants grew within one hundred feet of the nest, tens of thousands of years ago.

Pack rats collect virtually all the plants that existed in their area, so their middens make a good museum. The evidence shows that, on average, life zones were generally 2,500 feet lower during the ice ages—most modern plant species that existed during that time would have lived at an elevation roughly 2,500 feet below where they do today.

The impact of human life on the Grand Canyon can also be found amid its rocks and stones. These arrowheads found in the Grand Canyon are evidence of early Native Americans.

Scant evidence remains of the Archaic Native Americans who hunted and gathered in the Grand Canyon 3,000 to 4,000 years ago. Most intriguing of their few recovered artifacts are the folded and wrapped twig effigies of game animals these people hunted—deer and bighorn sheep predominantly. Scientists have found these split-twig figurines by the hundreds in the dark interiors of Grand Canyon caves. Several are currently displayed at the Tusayan Museum, located between Grand Canyon Village and Desert View.

A Sloth in the Rain: A Story

The details of the life of *Nothrotheriops* in this story are based on knowledge gained from studying the sloth's skeleton, preserved fur, and dung, all discovered within the Grand Canyon. It is also possible to extrapolate its basic habits and physiology by comparison with its nearest living relatives, the tree sloths of South America.

Twelve thousand years ago, the most recent ice age was coming to an end. It was late summer and rain was streaming down. The southwest monsoon had returned after an absence of almost 100,000 years. *Nothrotheriops*, the ground sloth, stood in the entrance of a cave that opened onto the Tonto Platform from the base of the Redwall cliff. A puddle was filling from a steady drip in the ceiling. *Nothrotheriops* looked out at the overcast day, cool after a hot summer. The sun had peeked out between thunderheads. The cave was a shelter, and its entrance room became a gigantic brood chamber for the local sloth population. Mountain goats sometimes shared the shelter, which felt warm in the winter, cool in the summer, and was always dim but comforting. The floor was soft with thousands of years of pungent dung.

An artist's view of a Shasta ground sloth, like the one described in this story

Nothrotheriops was old for a ground sloth, slower-moving even than most of its slow tribe. It was about eight feet long, including its sturdy tail, and it weighed 400 pounds. It was a herbivore. It was always chewing, waddling to find food, or sleeping.

Nothrotheriops, the largest animal within the Grand Canyon, was quite safe from predators. No bones of the saber-toothed cat have been found in the Grand Canyon, perhaps because the cat's chief prey, the mammoth and giant sloth, were too heavy to negotiate the steep terrain. *Nothrotheriops* had formidable claws on its front feet, capable of slashes that could send any cat running.

The rain poured down, bouncing from the stones and creating a rising mist. Across the Tonto Platform, pinyon pine (*Pinus edulis*) and juniper (*Juniperus osteosperma*) trees were winter-pruned to the exact height of *Nothrotheriops* standing on its back feet with its long neck stretched out. The underbrush, which the sloth also browsed, was ephedra, yucca, cacti, grasses, and numerous flowering shrubs and herbaceous plants. The sloth felt hungry, as it usually did while awake. Its primitive digestive system was relatively inefficient, requiring the animal to eat proportionally more forage than the advanced mammals of its world. This took time, and *Nothrotheriops* didn't have much of that, since it required many hours of sleep each day and night. So, walking on its front knuckles and the outside edges of its hind feet, *Nothrotheriops* ambled down the slope to eat.

Nothrotheriops stopped at a barberry bush (*Berberis fremontii*) that still had its dark blue berries. It ate the berries and all the youngest foliage it could find. The leaves were prickly and a little too tough for its few worn teeth. Perezia (*Acourtia wrightii*) and verbena (*Verbena macdougalii*) grew nearby. Using its prehensile lips, *Nothrotheriops* plucked up some of those before its attention turned to a nice ephedra bush, one of its favorites. The sloth didn't mind the rain as long as its body temperature was right—even modern sloths do not have an efficient temperature regulation system.

Was *Nothrotheriops*' inability to regulate its internal temperature the cause of its extinction as the climate warmed? Or was it the Clovis people, the Native Americans who first appear in the archaeological record at the end of the last ice age, with their long spears, their fire, and their advanced hunting technology? Perhaps both climate and humans can share the blame for *Nothrotheriops*' extinction, which occurred about 12,000 years ago. Meanwhile, *Nothrotheriops*, standing in the rain, suffering from advancing arthritis and feeling sleepy, returned to the cave for a nap. *Nothrotheriops* paused on the way for a long drink from a stream where fresh rainwater splashed over a rock into a pool. Then it entered the cave and settled down in the dung blanket. Paleontologists would find its bones in Rampart Cave, in the western Grand Canyon, 12,000 years later.

This skull of a *Nothrotheriops* was found in Rampart Cave in the Grand Canyon.

Epilogue: THE CONDOR RETURNS

The California condor, a bird that soared above the Grand Canyon thousands of years ago, is enjoying a rebirth.

At the close of the most recent Pleistocene ice age, about 11,000 years ago, there was a mass extinction of large American mammals. Some 70 percent of all species of animals weighing more than 100 pounds disappeared forever. With the extinction of this fauna, the carrion-eating California condor was left without its accustomed carcasses of big game. Little by little, the condors died out in the Grand Canyon area, or used their nine-and-a-half-foot wingspans to soar to the California coast. There, a reliable diet of beached cetaceans and stranded dead fish sustained a small population until modern times.

The world population of California condors had dropped to twenty-two birds by 1982, but the population is currently recovering, now numbering more than three hundred and increasing. During the summer of 2003, a mated pair of condors returned repeatedly to their Grand Canyon cave nest, carrying food—suggesting that a chick was inside. The condors had been raised in captivity and released near the canyon. Finally, on November 5, the chick, now as big as its parents, waddled to the edge and jumped off. It was the first flight of a wild-hatched condor since their near-extinction.

In another cave nearby, biologists had climbed down to an already-abandoned condor nest. Among fragments of modern eggs, they found 10,000-year-old California condor bones. After a long absence, the birds had returned to their ancient nesting site.

•   •   •   •

In this book we have explored the prehistoric life and environments of the Grand Canyon through almost 2 billion years, from primitive cyanobacteria to humans. This precious legacy is now carefully shelved in museum collections at the Grand Canyon and elsewhere. Today, the canyon's fossils are available for scientists to study, improving our knowledge of ancient life as no fossil hidden away in a private collection ever could.

It is a natural human impulse to covet a fossil and want to own it. But protecting the Grand Canyon's treasures demonstrates a loftier passion: to relinquish ownership in favor of stewardship. Learning more about the canyon heightens one's interest in protecting it. I hope this book has inspired you to further study and to extend kinship to all who, out of love for the Grand Canyon, are determined to adhere to Teddy Roosevelt's exhortation: "Leave it as it is."

GLOSSARY

acritarchs — spherical, cyst-like fossils

anaerobic — lacking oxygen

benthonic — describing bottom-dwelling animals

biostratigrapher — a scientist who studies the animals found at various strata in the fossil record

brachiopod — a type of shelled bivalve undersea creature

calcareous — describes structures that contain calcium such as shells, exoskeletons, and bones and function to support or protect an animal

chert — hard, silica-based sedimentary rock

dolomite — a rock that was originally limestone but to which magnesium-rich seawater filtered through the layers, changing the limestone to dolomite

fossilization — the process of turning animal or plant matter into fossils

functional morphologist — a paleontologist who discovers a fossil organism's physical capabilities by analyzing its internal structure

gneiss — a common metamorphic rock, similar in composition to granite but in which the mineral grains are aligned in distinct bands

ichnologist — a scientist who studies ancient tracks and trails

igneous — referring to a rock that forms when molten rock cools

Linnean hierarchy — the classification of animals and plants according to a system devised by Karl Linne (Linneaus)

metamorphic — any rock that forms from the alteration, through heat and pressure over time, of a pre-existing rock

midden — hardened accumulations, often found in caves and made by pack rats, of animal feces, urine, and discarded plant parts

nektonic — describing free-swimming animals

paleoecology — a study of ancient environments and their relationships with each other and with their environments

pluton — a body of igneous rock that forms when molten rock rises from the Earth's interior into the overlying crust and cools without erupting at the Earth's surface

regression — the event in which ocean waters retreat from large portions of a continental land mass, exposing it to erosion

sedimentary — referring to rocks that result from the accumulation and consolidation in layers of particles of rock debris or of dissolved rock material

strata — layers of rocks across broad areas

stratigraphy — the study of rock layers

stromatolites — fossils of ancient cyanobacteria that leave distinctive layered shapes in the rocks

unconformity — a gap in the geologic record, where one rock layer overlies another of substantially greater age; the gap may be due to nondeposition or to erosion of previously existing layers, or both

xerophytic — describing plants that are adapted to survival in drought conditions

SUGGESTED READING

Blakey, Ron, and Wayne Ranney. *Ancient Landscapes of the Colorado Plateau.* Grand Canyon, Ariz.: Grand Canyon Association, 2008.

Fortey, Richard. *Fossils: The History of Life.* New York: Sterling, 2009.

———. *Trilobite: Eyewitness to Evolution.* New York: Vintage, 2001.

Johnson, Kirk, and Richard Stucky. *Prehistoric Journey: A History of Life on Earth.* Golden, Colo.: Fulcrum, 2006.

Parker, Steve. *The World Encyclopedia of Fossils and Fossil Collection.* London, U.K.: Lorenz Books, 2007.

Price, L. Greer. *An Introduction to Grand Canyon Geology.* Grand Canyon, Ariz.: Grand Canyon Association, 1999.

Sadler, Christa. *Life in Stone: Fossils of the Colorado Plateau.* Grand Canyon, Ariz.: Grand Canyon Association, 2005.

Thompson, Ida. *National Audubon Society Field Guide to North American Fossils,* rev. ed. New York: Alfred A. Knopf, 1995.

PHOTOGRAPHY CREDITS

ILLUSTRATION CREDITS

INDEX

ABOUT THE AUTHOR

Dave Thayer operates a tour company at Grand Canyon National Park. He is one of the lucky group of enthusiasts who have spent their lives exploring and studying the Grand Canyon. He was geology instructor and science division chair at Yavapai Community College in Prescott, Arizona.

After moving to Tucson, he was Curator of Geology at the Arizona-Sonora Desert Museum. He appeared frequently on the television series *The Desert Speaks*, and more recently on a BBC program about geologic time in the Grand Canyon. Thayer was on the faculty of the Audubon Desert Institute and has taught for Elderhostel, and, currently, for the Grand Canyon Field Institute. Thayer is the author of *A Guide to Grand Canyon Geology along the Bright Angel Trail* (1986), the *Checklist of the Wildlife of the Grand Canyon* (2003), and the *Field Guide to Geology along the Bright Angel Trail* (2004). Thayer lives in Williams, Arizona, with his wife, Dora, who shares his love for all things Grand Canyon.